THE PRICE OF FREEDOM

Clayton Franklin Howell (1919-1942)
An American WWII Soldier in Bataan

DANA MOORE GRAY, PHD

TABLE OF CONTENT

DEDICATION

This book is dedicated to my uncle Clayton who was a true American hero, his mother and my grandmother Mary Margaret Neatherlin Howell who lived this book and spent countless hours seeking information and gathering family artifacts, and to my Aunt Margaret May (Bobbye) Howell Hoeme and my mother Charmagne LaUna Howell Moore who preserved the items their mother treasured.

FOREWORD

Treasured family heirlooms pass from one generation to the next, bringing rich insights into personal family history and ancestors' lives. My Aunt Bobbye left us a box of family photographs and documents including details about my Uncle Clayton's short life. The pictures, newspaper clippings, letters, and military documents demonstrate a mother's love and grief for a son who bravely paid the ultimate price for our freedom.

"He always thought he was coming home," said Clayton's sister Charmagne Howell Moore after reading letters from her brother included here. But PFC Clayton Howell didn't come home alive. He died at age 23 in the Philippines as a POW during WWII. His remains were sent by train to his parents, James Howell and Mary Margaret Neatherlin Howell.

He is one of countless American heroes whose stories deserve to be told to remind the living that freedom isn't free.

The author is grateful to Union County Leader newspaper publisher/editor Terry Martin for permission to include images of their newspaper clippings, the NM Bataan-Corregidor Memorial Foundation for their vital work to remember these heroes, the Defenders of the Bataan & Corregidor Memorial Society for documenting and sharing important history, and Marcus Griffin for compiling and editing the *Heroes of Bataan* book.

CHAPTER 1: CHILDHOOD

Clayton Franklin Howell was born January 20, 1918 in Coldwater, Comanche County, Kansas to James Howell (1898-1947) and Mary Margaret Neatherlin Howell (1899-1952). His paternal grandparents were John Howell (1859-1937) and America Jane Layton Howell (1854-1921). His maternal grandparents were James Franklin Neatherlin (1864-1919) and Emma Elizabeth McDonald Neatherlin (1875-1931. His grandparents were pioneers, the children of Texas Rangers and cattle drive bosses during post-Civil War cattle drives, and farmers and teachers settling the prairies of Kansas and Texas. Clayton was the first of nine children born to James and Mary who had a total of four sons and five daughters.

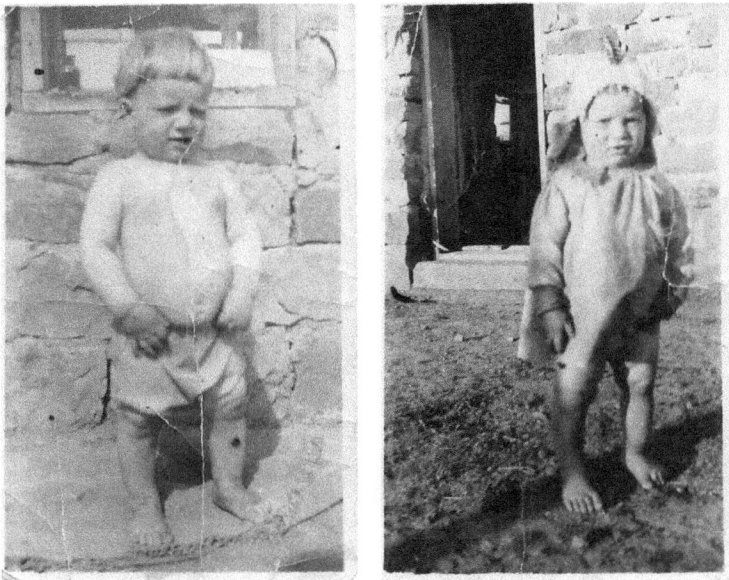

Writing on the back of the photo on the right is "Clayton in a little old dress. He is 'some fat boy.' He is only 9 days older than Hazel's baby but is twice as big and fat. Oh my!"

Clayton and his mother Mary Howell. Circa 1921 in Clayton, New Mexico.

Mary Neatherlin Howell and son Clayton, about 1919, New Mexico.
Inscription on back: "Don't I look with this short apron on and my hair isn't fixed up either."

Clayton and his mother.

EMMA McDONOLD NEATHERLIN
CLAYTON HOWELL--RAYMOND HOWELL about 1924

In the 1930 US census, Clayton was 11 years old, attending school, and living with his parents and four siblings in a home at 223 East Broadway in Clayton, Union County, New Mexico. He graduated from Clayton High School in 1937. In the 1940 US census, Clayton was 21 years old, single, and living with his parents and seven siblings at 408 Second Street in Clayton. He had completed four years of high school and was working as a "new worker."

On October 16, 1940, Clayton registered for the draft. His WWII draft card described him as 21, 5 foot 9 ½ inches tall, weighing 160 pounds, with a light complexion, brown hair, and gray eyes. His father James Howell was listed as next of kin. The card listed his address as 310 N. Second, Clayton, Union County, New Mexico and confirmed his birth date and location.

All four sons of James and Mary served in the military: Clayton Franklin Howell (1919-1942) joined the U. S. Army 200th Coast Artillery; Raymond James Howell (1922-1997) served in the Coast Guard during WWII; Andrew Mack Howell (1924-1949) was honorably discharged December 1945 from the US Navy Reserve as a Seaman First Class and gunner on the ship, then served in the US Army as a machinist until his untimely death on June 24, 1949; and PFC Moral Earnest (Hootie) Howell (1929-1969) who served in the US Army 363 Ordinance Ammo Company during the Korean War.

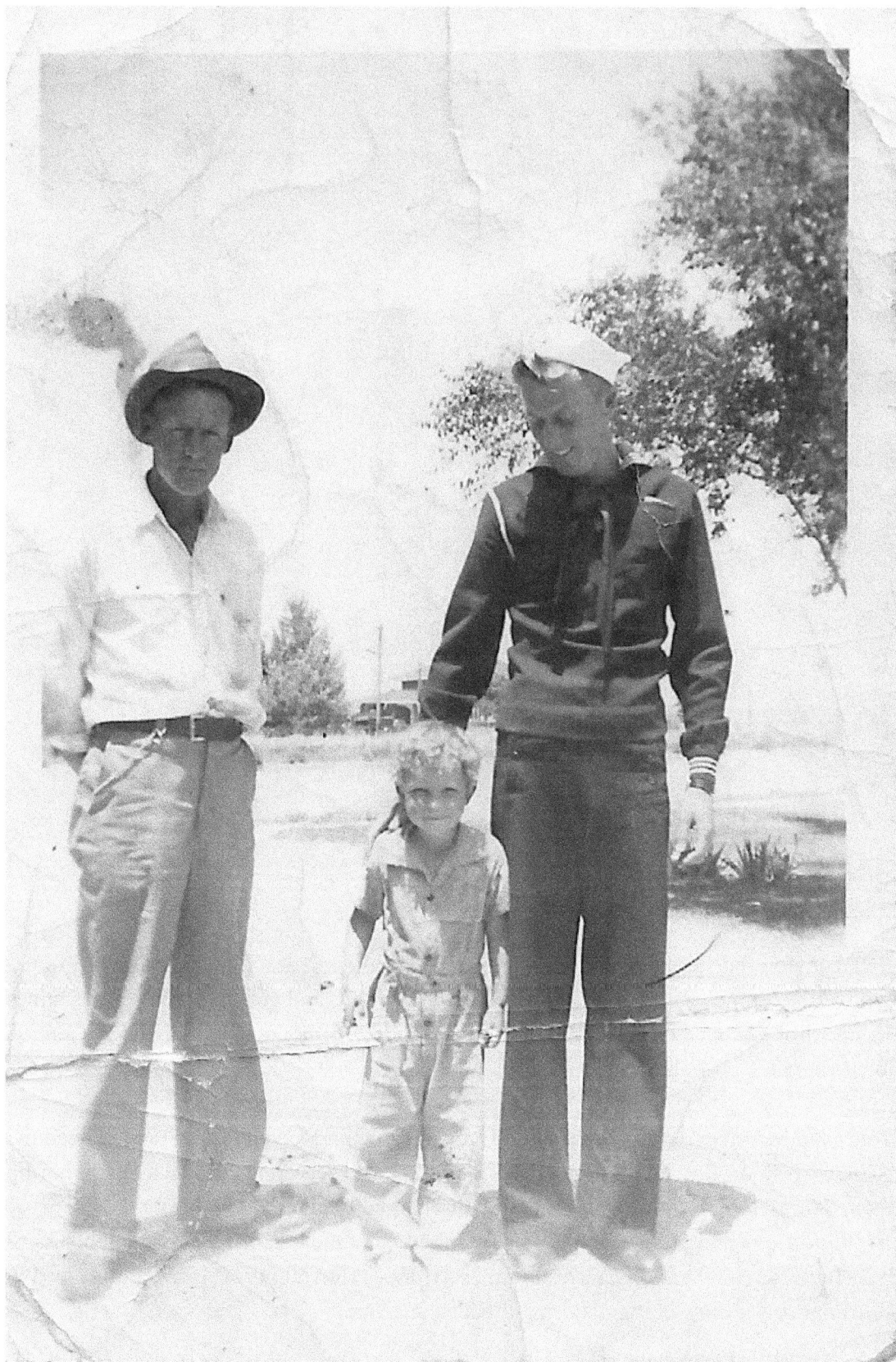

James Howell (left) and Raymond Howell, June 1943. Boy not identified.

Andrew Mack Howell and wife, about mid 1940s.

Moral Earnest "Hootie" Howell

Clayton's sisters (in order by age and height): Emma Lou, Margaret (Bobbye), Charmagne, Kay, and Sabra Howell. Circa 1945 in Clayton.

Howell children, circa 1970s. Left to right: Margaret (Bobbye), Charmagne, Raymond, Emma Lou, Kay, and Sabra.

CHAPTER 2: IN THE ARMY

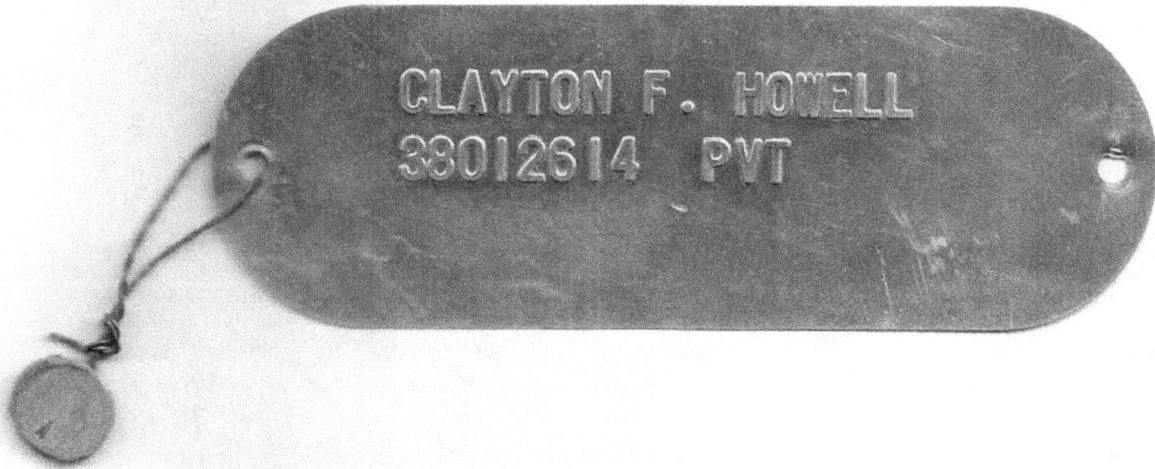

On April 4, 1941 in Santa Fe, New Mexico, Clayton enlisted with the U. S. Army and completed months of training in Ft. Bliss, Texas. There he joined other young soldiers from New Mexico to form the 200th Coast Artillery Regiment, A Battery. He was a private first class with serial number 38012614. After training, the 200th division (aka the Regiment) was deployed to the Philippines, traveling there via California and Hawaii.

https://www.bataanmemorialfoundationnm.org/

Note: Clayton is front row, second from left.

Written on back: Ready for chow. Date and location unknown.

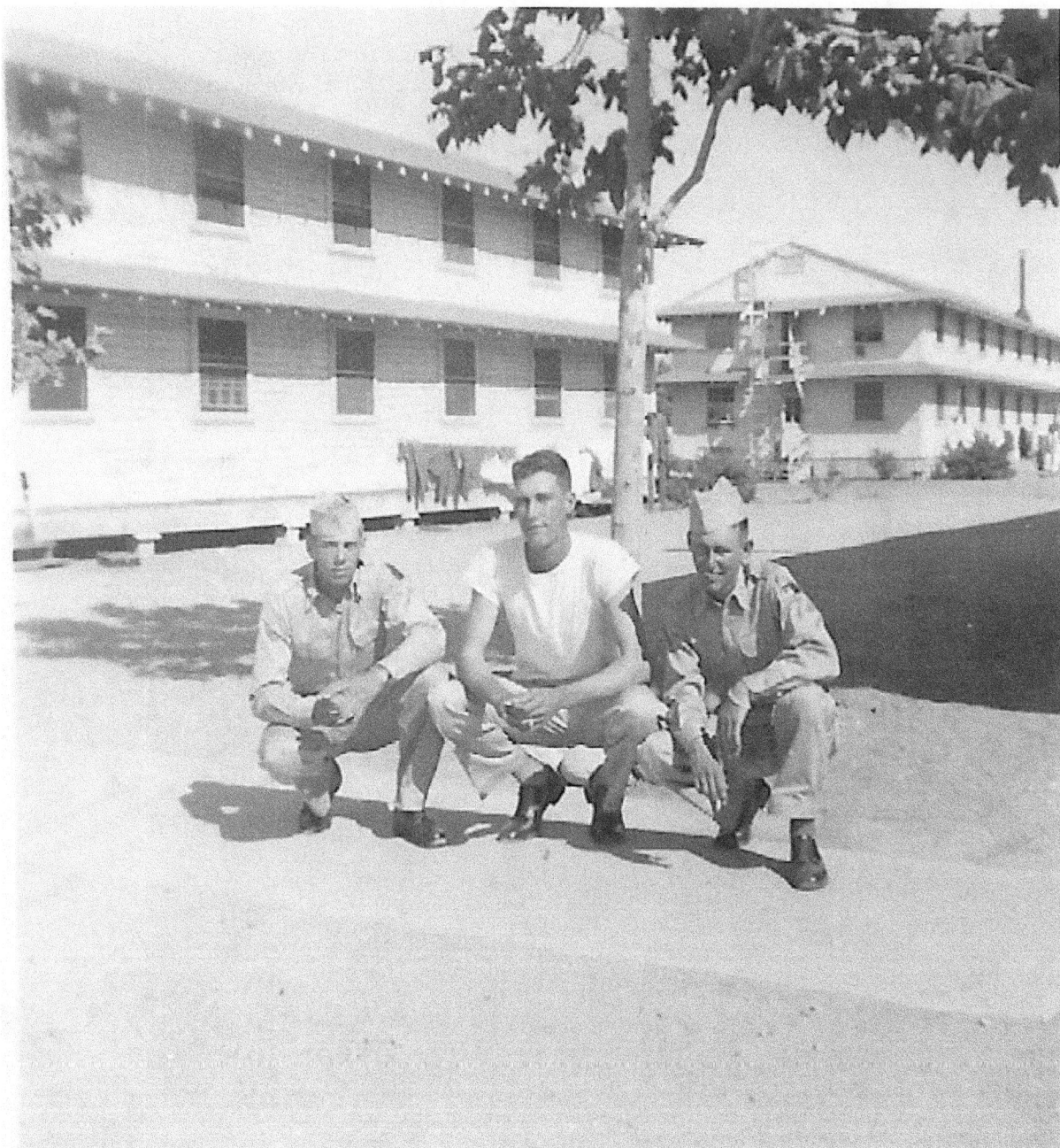

"Here is the three boys that stay hunger 24 hours a day." - Clayton Howell

"Don't laugh at these. They are some of my best friends." - Clayton Howell

After several months of strenuous training, the 200th (dubbed the Regiment) was officially named the best anti-aircraft Regiment of the US military. The unit was notified on August 17, 1941 that it had been chosen for an important overseas assignment (NM Bataan-Corregidor Memorial Foundation, n.d.). The unit was deployed, leaving Texas for California and then sailing to Honolulu and the Philippines.

The first stop was in California. On August 24, 1941, Clayton mailed a postcard of Quarantine Island, California. On the back he wrote that he was fine and thought his unit would sail the next day for Hawaii and then the Philippines. He and his unit left California on August 28, 1941 and sailed aboard the USS President Pierce to Honolulu, arriving on Sept. 3.

QUARANTINE STATION, ANGEL ISLAND, CALIF. 6128

Dear Mother, I am fine. It sure is nice out here. I think we sail tomorrow. I'm not so sure. Tell every body I said hello. I am at Angel Island now.
Love
Clayton

From Pfc Clayton Howell
200th CA AA Coy A

POST CARD

THIS SPACE FOR ADDRESS ONLY

PILTZ COMPANY, SAN FRANCISCO, CALIF.

Mrs. James Howell
11 Cherry St
Clayton
N. Mex.

The Union County Leader newspaper in Clayton, New Mexico published news that Clayon was stationed in Honolulu before moving on to the Philippines.

CLAYTON HOWELL ON WAY TO PHILLIPINES

Clayton Howell, who volunteer-ed for army duty April 2nd is now in the 200th CAAA Battery A and is stationed in Honolulu, Hawaii for the time being. They are on their way to the Phillipine Islands.

The U. S. S. President Pierce

U.S. Navy - U.S. Navy photo from Navsource.org, Public Domain,

https://commons.wikimedia.org/w/index.php?curid=27407443

Clayton wrote to his parents on September 2 and 3, 1941 and mailed the letters after arriving in Honolulu. In the letters, he mentioned his brothers Andy and Hoot (Moral Earnest), his Aunt Winnie (Winona Elizabeth Neatherlin Hecht, 1899-1980), and her son Clyde Benjamin Hecht (1920-2005). Hootie's sister Charmagne described him as ornery and teasing but not mean.

Dearest Folks,

We have been on the water since the 28th and are due in Honolulu some time in the morning. I signed up to go on a tour with the Chaplin to see the Island. I don't know if I will get to or not. From there we are sailing for the Philippines. It sure has been a tiresome trip so far but we still have about 10 to 15 days from Honolulu. We all go pretty well sea sick but most of us have our sea legs now. We are on the President Pierce. I don't expect I'll write very much over here probably once a month. It cost so much to send letters airmail and it would take an ordinary letter a month to get over there. I stayed at Angel Island for 3 days it sure is nice over there but its so hot here that we can scarcely breath. Tell dad and Andy they can do what they want to with the old Ford and tell Raymond I said hello. The Captain just asked me if I had a clean uniform and a dollar so I could go ashore at Honolulu I told him yes, He said "Well have you got $2 I would like to go to. I think I'll probably start to school over at the Philippines and learn radio as we won't use the telephone system. I bet old Hoot is mean as ever isn't he? If Aunt Winnie hears from Clyde have him to write to me once in a while. I'm not going to write to very many people. Hes somewhere in Washington in the air care. I just got back from 45 minute exercises and its about time to eat dinner. We get our cigarettes out here for 65 cents a carton everything is pretty cheap with the taxes all taken off. Well I'll close for now. Maybe get a chance to write more later. Love, Clayton

Sept 2 1941,
Honalula, Hawaii

Dearest Folks.

We have been on the water since the 28th and are due in Honalula some time in the morning. I signed up to go on a tour with the Chaplin to see the Island I don't know if I will get to or not. From there we are sailing for the Phillipines. It sure has been a tiresome trip so far. but we still have about 10 to 15 days from Honalula. We all got pretty well sea sick but most of us have our sea legs now. We are on the President Pierce. I don't expect ill write very much over here probably once a month. It cost so much to send letters air mail. and it would take an ordinary letter a month to get over there I stayed at angel Island for 3 days it sure is nice over there but its so hot here that we can scarcely breath. Tell dad and andy they can do what they want to with the old ford. and tell Raymond I said hello. The captain just ask me if I had a clean uniform and a dollar so I could go ashore at Honalula I told him yes. He said " Well have you got ½ I would like to go to. I think I'll probably start to school over at the Phillipines and learn Radio as we won't use the telephone system. I bet old Hoot is mean as ever. Isn't He? If Aunt Minnie hears from Clyde have him to write to me once in a while. I'm not going to write to very many people. I haven't heard from Fred Merilatt. He's somewhere in washington in the air core. I just got back from 45 minutes exercises. and its about time to eat dinner. We get our ciguettes out here for 65 cents a carton everything is pretty cheap with the taxes all taken off. Well I'll close for now. Maybe get a chance to write more later Love

Clayton.

Sept 3. Dear Mother, we docked at here about 5 this morning am O.K. Getting ready to go out on shore leave. I'll write as soon as I get in them the Phillippines

Sept. 3. Dear Mother, we docked at here about 5 this morning. Am O.K. Getting ready to go out on shore leave. I'll write as soon as I get in the Philippines.

By October 12, 1941, Clayton had been stationed in Pampanga, Philippines, a province in Central Luzon bordering Bataan, Bulacan, and Zambales. From there he wrote his parents two letters, mentioning his younger brothers Raymond, Andy, and Hootie (aka Moral Earnest).

Pampanga Oct. 12

Dear Folks.

I recd your letter a couple of days ago and this being Sunday I will have time to write a few lines. I owe several letters that I must try and write this eve. There isn't much news I am well. We haven't got to go hardly anywhere yet. There are several small towns around here. Not much in them but natives. I have quite a lot of pictures of them and scenery of the country. This is really a Paradise over here. But it isn't much good to work in as a little exercise sure gets you down. All the outfits over here are really good. The air field is just a block away. Im glad Raymond got a good job. Maybe he can get a chance to stay with it. Its rumored here that we may get ~~to go back in Dec~~. If we do this will be a nice trip. Its winter over here now and its so hot that it is pitiful never gets below 70 degrees at nite. Everything is green. A fellow can go out anywhere and pick bananas coconuts off the trees. I have lots of friends here so its not so bad. I must write Clyde and Fred merriatt. Also several others. I would like to go to manilla and get some souvineers and send but don't know when I can. and when I get hold of enough money. Ill try and send you some for stamp money later on. I am to start school in the near future. I figure on taking up electrical engineering and electricty we have already filled out the forms. I also have been transferred to a new radio locater. I like it fine. I hope to learn something that will be of value to me when I get out. Well I had better sign

off and write some later on. I'll try and write at
least two letters air mail a month I believe that
will be plenty as there is hardly any news over
here

love
Clayton

Dear Dad,
I hope this finds you well. I haven't written much
on account of there isn't hardly any news. I'll try
and send some stamps and maybe later on ill go to
Manilla and buy some various types there. Tell
Clene I said hello and im still just a good republican
as ever. You needn't worry about me drinking anything
over here. The weather is so hot that almost ever
one that drinks goes crazy. They send them back on
Section 8. I wouldn't want to get out that way. I
haven't drink anything but a few beers in 4 months.
I hope Raymond can get a good job with the
Co & keep it. I don't know what ill do when I get
out. Tell Andy I saw a little monkey up in the
jungle last Sunday that looks just like Skorbie. I'll
catch one and bring it home if they'll let me. Well Ill
close for time being

as ever
Clayton

Dear Folks, I rec'd your letter a couple of days ago and this being Sunday I will have time to write a few lines. I owe several letters that I must try and write this one. There isn't much news. I am well. We haven't got to go hardly anywhere yet, There are several small towns around here. Not much in them but natives. I have quite a lot of pictures of them and scenery of the country. This is really a paradise over here. But it isn't much good to work in as a little exercise sure gets you down. All the outfits over here are really good. The air field is just a block away. I'm glad Raymond got a good job. Maybe he can get a chance to stay with it. It's rumored here that we may get to go back in Dec. If we do this will be a nice trip. Its winter over here now and its so hot that it is pitiful never gets below 70 degrees at nite. Everything is green. A fellow can go out anywhere and pick bananas or coconuts off the trees, I have lots of friends here so its not so bad. I must write Clyde and Fred Merilatt. Also several others. I would like to go to Manila and get some souvenirs and send but don't know when I can and when I get hold of enough money. I'll try and send you some for stamp money later on. I am to start school in the near future. I figure on taking up electrical engineering and electricity. We have already filled out the farms. I also have been transferred to a new radio locator. I like it fine. I hope to learn something that will be of value to me when I get out. Well I had better sign off and write some later on. I'll try and write at least two letters airmail a month I believe that will be plenty as there is hardly any news over here. Love, Clayton.

Dear Dad, I hope this finds you well. I haven't written much on account of there isn't hardly any news. I'll try and send some stamps and maybe later on I'll go to Manila and buy some various types there. Tell Cline I said hello and I'm still just a good a republican as ever. You needn't worry about me drinking anything over here. The weather is so hot that almost every one that drinks goes crazy. They send them back on Section 8. I wouldn't want to get out that way. I haven't drank anything but a few beers in 4 months. I hope Raymond can get a good job with the Co.. and keep it. I don't know what I'll do when I get out. Tell Andy I saw a little monkey up in the jungle last Sunday that looks just like Hootie. I'll catch one and bring it home if they'll let me. I'll close for time being. As ever, Clayton.

Clayton again wrote to his mother a few days later on October 15.

Dearest Mother,

I should have mailed this but haven't had time until today. I have been going to school and it sure gives a fellow a headache over here. Radio and electricity is something I never had studied so I have to study a little harder than the most of them in order to learn it. It is rumored here that if nothing happens will head back to the states by xmas. Part of the men over 28 are going as soon as they can catch a transport. There isn't much news over here. So I'll close as I have to take a shower and shave. We have to take at least 2 baths a day. Tell them all hello. Love, Clayton.

Close to Christmas of 1941, Clayton mailed to his parents a Christmas card printed on rice paper, folded, with a postcard image attached to rice paper. The image on the front was part of a postcard.

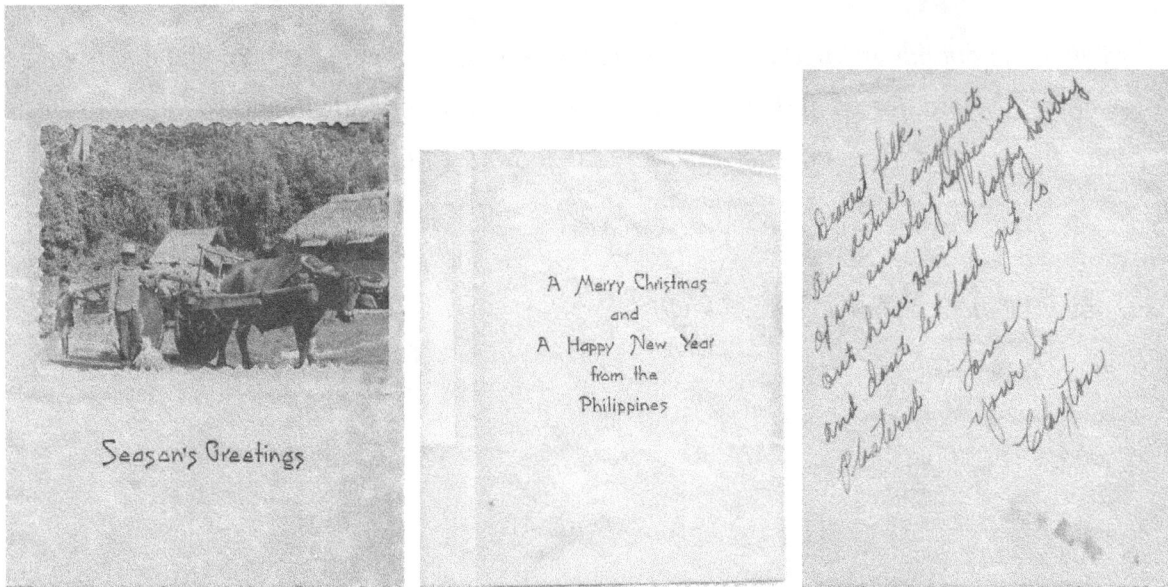

Then everything changed. Japan attacked Pearl Harbor in Honolulu, Hawaii on December 7, 1941. Just hours later, Japanese forces bombed Clark Field in the Philippine Islands (technically on December 8, 1941 because of time zone differences). Clayton's unit - New Mexico's 200th Coast Artillery anti-aircraft regiment - was reportedly the first to return fire in the Philippines. The unit was split that night to provide anti-aircraft protection for Manila, and formed the 515th Coast Artillery, the first battle-born unit of WWII, according to the Bataan Memorial Foundation. Clayton stayed with the 200th.

The NM Bataan-Corregidor Memorial Foundation (n.d.b) wrote "At 12:35 hours, 8 December, Manila time, Japanese bombers, flying at 23,000 feet and accompanied by strafing planes, made their appearance and the war was on. The 200th could not, with powder train fuses effective only to about 20,000 feet, do much damage to the high altitude bombers. The men dished out whatever they could and stood up well under these unfavorable and unequal conditions. When the smoke from the muzzles cleared away, five enemy planes had been shot down and two men of the outfit had lost their lives" (para. 6).

Fighting intensified near Clark Field and Manila over the next two weeks and the decision was made for US troops to withdraw to Bataan. The 200th covered the retreat of the Northern Luzon Force into Bataan while the 515th did the same for the South Luzon Force. Both units protected bridges allowing forces to retreat to Bataan, completing their missions (NM Bataan-Corregidor Memorial Foundation, n.d.b).

On December 28, 1941, Clayton sent his parents a telegram stating he was fine and sending holiday greetings. The Union County Leader newspaper published an article about the wire, noting Clayton was at Ft. Stotsenburg near Manila in the Philippine Islands. The assignment was to protect Fort Stotsenberg (NM Bataan-Corregidor Memorial Foundation, n.d.).

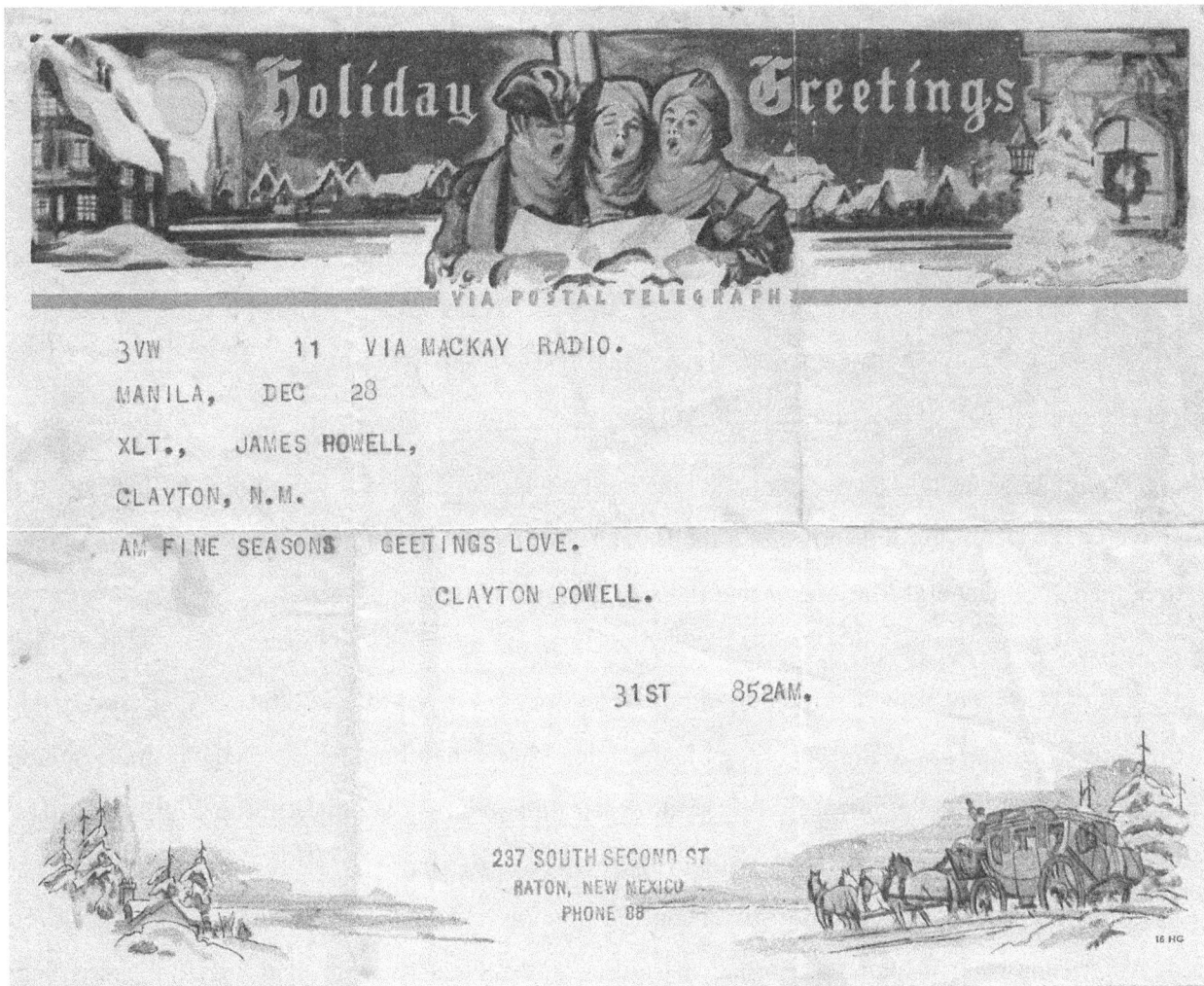

CLAYTON HOWELL SAFE

Mr. and Mrs. James Howell of Clayton were overjoyed to receive a wire from their son, Clayton, last week, stating that he was safe and well at Ft. Stotsenburg, near Manila, in the Phillipine Islands. The wire was sent Dec. 28, according to Mrs. Howell who added that her son had been in that area since last October. He is with the 200th Anti Aircraft division, which appears to be New Mexico's own group. Clayton, who is a graduate of the Clayton High School, class of 1937, joined up with the Army last April as a volunt...

The Union County Leader newspaper published news that Clayton's parents were notified February 14, 1942 that Clayton was not injured and safe in the Bataan Province of the Philippines. At the time, Bataan had been defended by General MacArthur and then by General Wainwright.

Clayton's sister Charmagne remembered their mother continually prayed for the safety of Clayton and his Regiment while her father paced in front of the big console radio in their home that was the primary source of their news, cursing the government for not sending reinforcements or supplies to the 200th Regiment. Recall that the Great Depression had just ended (in 1941). Out of necessity some families grew closer during those difficult years of economic strife, sharing precious resources. Television had not yet been invented. Americans received news via broadcast radio, newsreels shown in local movie theatres, and local newspapers. The small community of Clayton, New Mexico was closely knit and anxiously awaited any news of the welfare of their hometown soldiers.

Fighting continued in the area for the next few months and American troops suffered severely from malaria, dysentery, and food shortages. The suffering and starvation led to low morale. Japanese troops received reinforcements and began their drive down the peninsula on April 3, 1942, fueling intense fighting until they broke through allied lines on April 7 (NM Bataan-Corregidor Memorial Foundation, n.d.).

The American Defenders of Bataan & Corregidor Memorial Society (n.d.) documented the torture and atrocities suffered by these soldiers. They wrote "The first group of men to arrive at Mukden numbered about 1500 and came from 2 different areas. The 1400 Americans sent to Mukden were part of the force that was responsible for delaying the fall of the Philippines and slowing down Japan's timetables for conquest of the Pacific. The soldiers of Bataan had survived horrendous battles on Luzon as they fought fresh and well supplied Japanese forces. They used mostly WW1 ammo and weapons that often failed. Adequate food supplies sat in warehouses and on the docks of Manila, so, as they shared their food with the Filipinos, they battled the Japanese, numerous tropical diseases, on half, then quarter rations. Sailors whose ships were lost and airmen whose planes were lost got quick infantry training with no live ammo. They fought fiercely despite suffering from a variety of tropical diseases and starvation" (para. 1-4).

The battle for the Bataan peninsula of Philippine Islands was lost on April 9. The 200th and 515th fought for four months and shot down 86 confirmed enemy aircraft before surrendering, overwhelmed by starvation and disease. The U. S. surrendered Bataan on April 9, 1942 and Corregidor on May 6, 1942. The 70,000 captured American and Filipino soldiers, many of whom were starving and injured, became Japanese prisoners of war (POW). The Japanese forced their prisoners to march 63 miles from Mariveles to San Fernando to a prison camp now called Camp O'Donnell (Wikipedia, n.d.a;).

The brutal walk in April of 1942 became famously known as the Death March of Bataan because of the horrors suffered by the prisoners. En route, 10,000 died and others escaped into the jungle (Britannica, n.d.). History documents the horrific torture and abuses were the "most inhumane known to mankind as prisoners of war," noting that nearly half of the 1800+ soldiers originally deployed there died in torturous Japanese prison camps or Hell Ships (Bataan Memorial, n.d.). The American Defenders of Bataan and Corregidor Memorial Society described the torturous Death March. ". . . the Japanese marched the remaining soldiers up to 70 miles in what is called the Bataan Death March. Men who could not keep up were killed. They were shot, disemboweled, beheaded, or bayoneted if they tried to get water or food. Some were run over by trucks and left in the road. They were left in the hot sun for hours to "rest." Local people, who tried to give them food or water, were killed for their kindness"(para. 4).

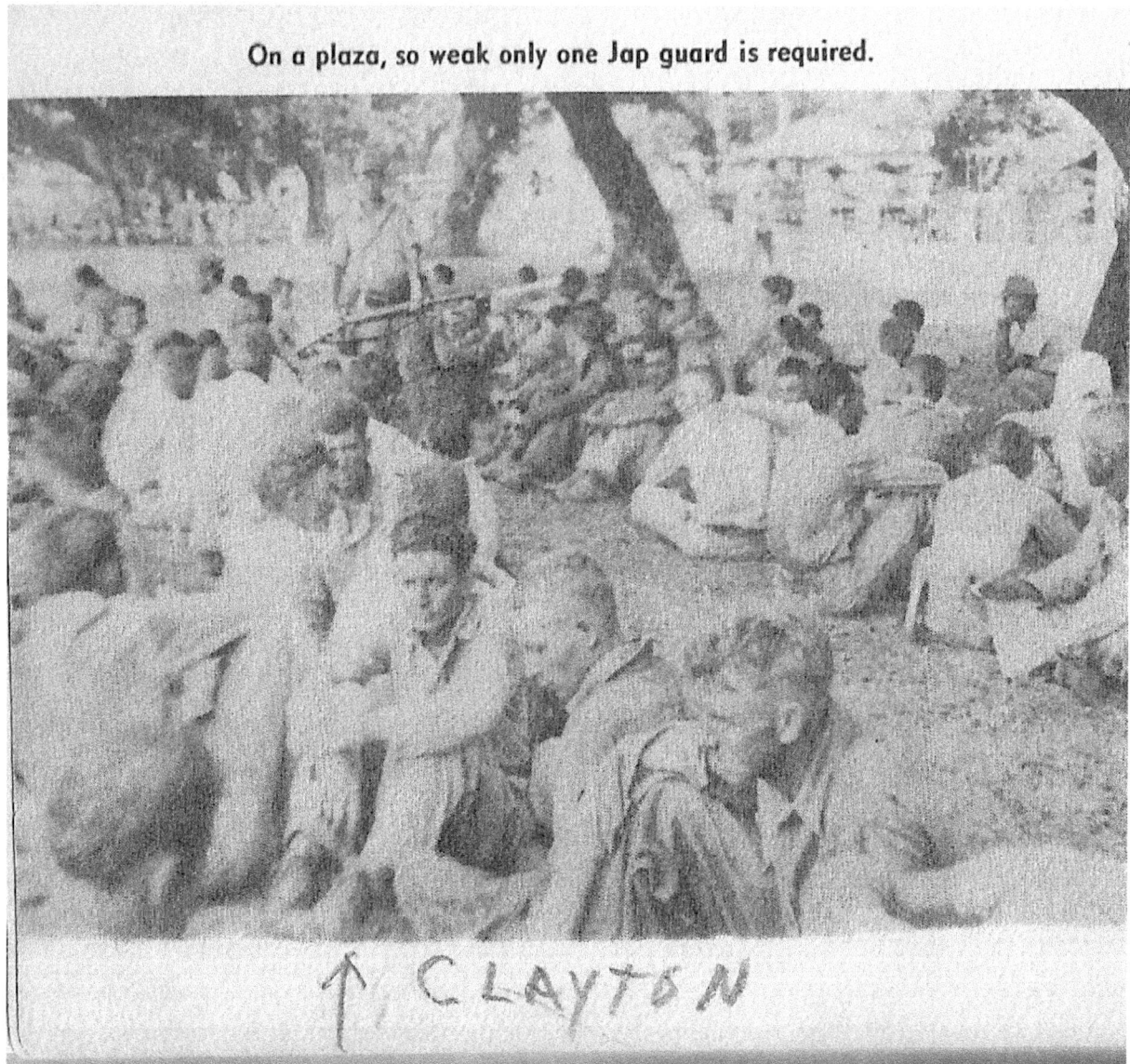

On a plaza, so weak only one Jap guard is required.

↑ CLAYTON

Griffin (n.d.) compiled a book about Bataan and included several photographs found with a dead Japanese soldier. The photo above was one of several images of American soldiers during the Death March. In her copy of Griffin's book, Clayton's mother identified her son in the photo above, writing his name with the arrow. Image: (Griffin, n.d.).

The Union County Leader newspaper published articles about local soldiers taken as prisoners of war.

LOCAL BOYS ARE JAP PRISONERS

Mr. and Mrs. Frank Henry were officially notified Friday that their son, Boyd N. Henry, was listed as a prisoner of war of the Japanese Government in the Philippines. Friday evening Mr. and Mrs. James Howell received a similar notification in regard to their son, Clayton F. Howell. Ben Williams, son of Mr. and Mrs. Ben Williams of Espanola, former residents of Clayton is also listed as a prisoner. This brings the number to five Union County boys, who were with the 200th C. A. who are listed as prisoners. The parents are glad to know that their sons are alive, and all hope that they will come home, in the near future.

Prisoner In Philippines

Mrs. James Howell received telegram from the War Department last Friday, that her son, Clayton, is a prisoner of the Japanese in the Philippines.

Clayton has been missing in action since the fall of Bataan. The last letter Mrs. Howell received from him was about ten months ago. Mrs. Howell was very proud to know that her son was still alive.

The Union County Leader newspaper published an article after Clayton's mother was notified May 20, 1942 that her son had been taken prisoner.

MRS. HOWELL GETS LETTER REGARDING SON ON PACIFIC

Clayton Howell, son of Mr. and Mrs. James Howell of this city is very likely a prisoner of the Japs, since he was with the 200th Coast Artillery, in the Philippines when the Japs first attacked the islands last December.

The following letter received by Mrs. Howell from Senator Chavez regarding the whereabouts and welfare of Clayton stresses the fact that her boy is apparently a prisoner of war.

May 20, 1942

Mrs. James Howell,
11 Cherry St.
Clayton, New Mexico,
Dear Mrs. Howell:

Some letters came out of the Philippines by plane, from those men who were near the field or base of operations. I assume your son was, hence, he could write you. The letters came out of Bataan, I am advised.

The name of your son was not on my list, nor was it on the casualty list of the War Department . Hence, you may assume he was interned by the Japanese prier to April 9. Prisoners are being accorded decent treatment, I am informed.

Clayton survived the infamous Death March of Bataan in April, 1942 and arrived in what is now called Camp O'Donnell (Wikipedia, n.d.a). After June 1942, Clayton was moved to Cabanatuan POW camp (Jackson, 1946) located in Nueva Ecija Province, Philippines. In early November 1942 he was moved to Hoten Mukden POW Camp in Manchuria (Mansell, 2005) where he died that month.

After Clayton was taken as a POW, his mother Mary reached out to countless people for any news of her son and his unit. Some wrote back over the next few months and years. Charmagne recalls one soldier visiting after the war to share what little information he had.

Clayton's mother received a thoughtful, two-page letter on August 12, 1942 from 1st Lt. A. N. C. Lucy I. Wilson, a WWII nurse. She did not know Clayton but took the time to share what information she had and to provide some comfort to the family.

August 12, 1942

My dear Mrs. Howell,

. I am very sorry I never had the opportunity of meeting your
son, Clayton Howell. Due to the fact that I had been in the
Philippines only six weeks and was stationed at Ft. McKinley before the war started, I had met very few people. I worked in
surgery in Bataan where my patients were under anesthetic so I
never got to know any by caring for them, as I never came in con-
tact with them any more after leaving surgery. The boys were on
duty 24 hours a day every day and never came down to the hospital
unless ill or injured. I was on Corregidor for only about three
weeks and therefore had not met many people there.

I realize that the relatives of people left behind will use
every possible chance to find out every little scrap of information they can, and I am willing to do anything I can to help. I
have written all the mothers of the boys I knew and have received
a great many letters like yours. I am taking the name and information from the letters I receive and sending it to each of the
girls in my group. If you do not receive an answer from any of
them, you will know they knew nothing about your son. I am sorry
I have been so long in answering your letter.

Some people are under the impression that their relatives
were in Cebu or Australia, which is usually wrong. A few were
able to send some mail or radiograms which were sent from there,
but the people themselves were not able to escape or their relatives would have heard from them again by now. Not everyone was
able to get mail out and it was impossible to get any in to us
after the war started and it was, therefore, returned to the
writer.

If you did not receive notice that your son had been injured up until the surrender, then you have good reason to believe that he is all right, except for the possibility of in-
jury during the last few days of battle and of that no one will
be able to give you any information except that released by
Japan through Red Cross, if and when they ever consent.

We presume that the people on Bataan and Corregidor at the
time of surrender to be prisoners of the Japs. I do not know
whether or not they are keeping them prisoners on the Islands
or have taken them to Japan. I have not heard of them being
moved anywhere.

As to how they are being treated no one knows and will not until released by Japanese through Red Cross. As soon as the Japs agree, the Red Cross will notify everyone as soon as possible without your having to make inquiry. Naturally, we tried very hard to find out how our prisoners were treated because we thought eventually we might be prisoners, but it was impossible.

As far as I know my group was the last to escape Corregidor—48 hours before it fell. I like to think they are being treated as nicely as the Japs are able to—why believe the bad? Faith and Hope must be our motto just as it was of the boys. I have many dear friends left there from whom I, too, will be so glad to hear.

Thank you for your kind wishes, and I pray for you and yours. There are no braver heroes or better fighters in the world than those who fought on Bataan and Corregidor. I have no words to describe the spirit and high morale of those. To me they are unequaled.

Sincerely yours,

Lucy I. Wilson

Lucy I. Wilson
1st Lt. A. N. C.

Mrs. James Howell
417 Cherry Street
Clayton, New Mexico

Timeline

- October 16, 1940 - Clayton, New Mexico - Clayton registered for WWII draft
- April 4, 1941 - Santa Fe, New Mexico, enlisted with U. S. Army
- April - August, 1941 - training at Ft. Bliss, Texas
- August 24, 1941 - Quarantine Island, California
- September 3, 1941 - Honolulu, Hawaii
- October 12, 1941 - stationed Pampanga, Philippines (near Manila)
- December 8, 1941 - Fort Santiago, Philippines
- December 8, 1941 - Japan bombed Clark Air Field in Philippines
- December 28, 1941 - stationed at Ft. Stotsenburg near Manila, Philippines
- Spring 1942 - Stationed in Bataan, Philippines
- April 9, 1942 - U. S. surrendered Bataan, Philippines
- April, 1942 - survived Death March of Bataan, 63 miles from Mariveles to San Fernando prison camp now called Camp O'Donnell, Philippines
- June 1942 - moved to Cabanatuan POW camp in Nueva Ecija Province, Philippines
- Early November 1942 - moved to Hoten Mukden POW Camp in Manchuria, Shenyang, China
- November 17, 1942 - died in Hoten Mukden POW Camp in Manchuria

CHAPTER 3: AFTER HIS DEATH

Military documents note November 17, 1942 was Clayton's last report date, and that is considered his date of death. He died as a POW in Hoten POW Camp (Mukden) Manchuria 42-123, an area now part of China. The cause of death has been reported as "disease." One family member thought the cause of death was dysentery and another said diphtheria. The Japanese did document in 1943 inoculating prisoners with various diseases including cholera.

The National Archives lists Clayton's name among the WWII Army and Army Air Force Casualties (n.d.).

TORRANCE COUNTY

Name	Number	Rank	Casualty
BARNES BULEN	38012185	PVT	KIA
BLANCETT LEONARD O	38158190	S SG	KIA
BOLTON SHILLEY L	38012033	PVT	DNB
BRASWELL ARNOLD O	18122314	PVT	DNB
CAREY DANIEL W	18122019	PVT	DOW
CAST ELZIE L	38012127	CPL	DNB
CAST VIRGIL A	38014905	SGT	KIA
CHAVEZ ABEL	38352822	PFC	KIA
GARCIA JOSE V	38159076	PVT	KIA
GARCIA LEONIDES L	38012182	PVT	KIA
GARCIA SANTIAGO T	38350822	PFC	DNB
HARDY EDWARD L	38584048	PFC	KIA
HODGIN RALPH L	38500331	PVT	KIA
HOLLEMAN NOEL A	18120744	S SG	DNB
HUGHES ANDREW J	O-663916	CAPT	FOD
LARRANAGA AMBROCIO M	38581288	PFC	KIA
LERMA JUAN S	38012190	PVT	DNB
LUCERO AMADO	38167504	CPL	KIA
LUNA ANDALECIO C	38581308	PVT	DOW
LUNA CANDIDO	38012485	PVT	KIA
MADRIL JOAQUIN	38168187	PFC	KIA
MAES ARTHUR	38580727	PFC	KIA
MALDONADO MIRAMON	38012183	PVT	DNB
MARTINEZ ADOLFO	38350144	PVT	DOW
MAYFIELD ARBIE J	38014099	PFC	KIA
MC NATH WILLIAM E	38166658	S SG	FOD
MOSELEY JUAN	38012473	PVT	KIA
OTERO TRINIDAD F	38012192	PVT	KIA
RODGERS MORTON S	38384862	TEC5	KIA
SALAZAR SANTIAGO C	39165230	PFC	KIA
SANCHEZ JUAN J	30011195	PVT	DNB
THOMAS BILLY J	20842590	PFC	DNB
VIJIL SAN J	38168180	PVT	DOW

UNION COUNTY

Name	Number	Rank	Casualty
ALARID JOHN L	38504059	S SG	KIA
ARLEDGE GARRETT M	38012197	PVT	FOD
ATCHLEY WILLIAM T J	38164546	S SG	FOD
EARP ROBERT R	38128770	SGT	FOD
ESPINOSA DELFIN G	38504067	PFC	KIA
GURULE FRANK	38352590	PFC	KIA
GURULE JUAN F	38011203	PVT	DNB
HENRY BOYD N	38012192	PVT	KIA
HOWELL CLAYTON F	38012614	PFC	DNB
JACOBS LOREN N	18121819	CPL	DNB
JIRON ARCHIE C	38166841	PFC	KIA
KUHNS WILLIAM C	6953615	PFC	DNB
LANE KENNETH L	38012613	PVT	DNB
LEIGHTON KENNETH E	38070431	PFC	KIA
MARTINEZ JUAN C	38010072	PFC	KIA
MARTINEZ MAX	38012870	S SG	KIA
MC ADAMS WILLIAM B	O-672828	1 LT	FOD
MILLIKEN GEORGE J	38012189	TEC5	DNB
NICKLAS J LAVERNE	O-327455	CAPT	KIA
ORTIZ BILLIE	38012434	PVT	DNB
PITMAN LEO	38012436	PFC	DNB
RADOSLOVICH C A	01385517	2 LT	KIA
RANKIN DAZE D	38580796	PVT	DOW
RIGDON GLENN N	38680797	SGT	KIA
ROSS NOEL D	18070140	S SG	KIA
SALAS JOE A	38169658	PVT	DNB
SMITH FAUAD J	38013678	T SG	KIA
TAYLOR LUIS	38012196	PVT	DNB
TRUJILLO TONY	38352588	PFC	KIA
TRYON DICK	6956166	SGT	KIA
VIGIL ALFREDO F	38581090	PVT	KIA
83 05Y			

VALENCIA COUNTY

Name	Number	Rank	Casualty
ANTONIO JACK	18068406	TEC4	KIA
ARAGON MARTIN NMI	38104843	TEC5	KIA
ARAGON VICTOR R	38193115	PVT	DNB
ARMIJO ANTONIO B	38010038	PFC	KIA
BACA NARCISCO B	38503971	PVT	KIA
BACA RAMON R	38504409	PFC	KIA
BOQUS LAURICE A	38122329	PFC	KIA
BUDENHOLZER FRANK A	O-361242	2 LT	DNB
CANDELARIA SOLOMON M	38120337	PVT	KIA
CASTILLO FILIMON C	38013440	PVT	DNB
CHAVEZ GRANT M	80841681	PFC	KIA
CHAVEZ RONOLO	38180375	PFC	KIA
CHAVEZ RUBEN S	38352924	PFC	KIA
CHAVEZ SAN J	38123794	PFC	KIA
CHAVEZ SATURNINO JR	38035297	T SG	KIA
COFFIN HOWARD	18016957	S SG	KIA
CORDOVA EDWARD	38011206	PVT	DNB
DALRYMPLE EVAN S	37383034	TEC4	KIA
DELAY JACK A	18119914	TEC5	DNB
DUNG EDDIE A	38168130	PVT	DNB
EVANS ROBERT E	O-433033	1 LT	KIA
FERNANDO PAUL	38013037	TEC5	DNB
FREEMAN LAURENCE R	38167419	S SG	KIA
GALLEGOS HERMAN	38352851	PFC	KIA
GARCIA ENRIQUEZ	38013199	PVT	DNB
GARCIA EUTIMIO C	38168742	PFC	DNB
GILL ERNEST E	T-120649	FL O	DNB
GUERRERA PAUL C	38011817	PVT	DNB
HICKS SAMUEL M	38167395	S SG	KIA
HOLCOMB JESSE C	38012594	PVT	DNB
HUGHES OSCAR J	38122357	PVT	KIA
JARAMILLO LEONOR	38167844	PFC	DOW
JOHNSON FRANCIS J	38352358	PFC	DOW
JONES FRANK G JR	38011211	CPL	DNB
KASERO ANTONIO	38012806	PVT	KIA
LAMANCE ELMER D	18058450	PFC	DNB
LEEDS FLOYD C	38132218	S SG	DNB
LENTE SEPERINO	38012819	PVT	KIA
LOPEZ BENITO	38100659	PFC	KIA
LOPEZ BILLY	6955668	S SG	KIA
LOVATO EUSEBIO Z	38168785	PFC	KIA
LUCERO FRANK D	38011747	PFC	KIA
LUCERO GREGORIO	38169976	PFC	DOW
MARTINEZ JEREMIAS G	38018801	PVT	DNB
MARTINEZ MANUEL J	38348445	PVT	DNB
MATKOVICH CHARLIE P	38123744	PFC	KIA
MC CRACKEN RUSSELL L	38584289	SGT	KIA
MOLINA REMIGIO	38348008	PFC	KIA
MORRIS JIMMIE	38011256	PVT	DNB
MUNSON LLOYD W	38012598	PVT	KIA
OLEE CHARLES M	38012809	SGT	DNB
PACHECO MARIANO A	38168762	PFC	KIA
PADILLA ALONZO M	38384512	PVT	KIA
PERALTA LOUIS	38167383	PFC	DOW
PEREA CLOVIS	38010647	PFC	KIA
RODRIGUEZ SANTIAGO C	38103912	PFC	KIA
ROMERO CLAUDIO	38012597	PVT	DNB
ROMERO POCH	38010296	T SG	KIA
ROMERO FRANK	38011856	PVT	DNB
ROMERO JOHN M	38503970	PFC	KIA
ROMERO JUAN V	38011772	PFC	KIA
SAIS ARTURO J	38583116	PFC	KIA
SAIS SANTIAGO S	38011266	PVT	KIA
SANCHEZ ADELARDO I	38012804	CPL	DNB
SANCHEZ FRANK J	38012810	PFC	DNB
SANCHEZ JUAN D	38169988	PVT	DNB
SARRACINO FRANK B	38012435	T SG	KIA
TAFOYA GRABIEL	38012227	PVT	DNB
TAFOYA JESUS M	38012224	PVT	DNB
TAFOYA JOSE M	38012213	PVT	KIA
TAFOYA MARTIN A	38012195	CPL	DNB
TAFOYA MARTIN J	38012205	PVT	DNB
TAFOYA RUPERT C	38010961	CPL	KIA
TORRES PEDRO	38167335	PVT	KIA
TORREZ JOSE	38352925	PFC	DOW
TRUJILLO EPIFANIO C	38012348	PFC	DNB
TRUJILLO FLORENIO	38073395	CPL	KIA
TRUJILLO JESUS P	38012198	PVT	DNB
TURRIETA JOSE E	38358031	PVT	KIA
ULIVARRI DANIEL B	38072396	PFC	KIA
VAN TINE DONALD R	38705535	PVT	DNB
VIGIL DOROTEO G	38352555	PVT	KIA

STATE AT LARGE

Name	Number	Rank	Casualty
BAYLEY WILLIAM C	01283710	2 LT	KIA
CAROTHERS SAM E	O-319083	CAPT	KIA
COTTEN CHARLES M	18063297	T SG	FOD
DAVIDSON JAMES L	01320609	1 LT	KIA
FAIRBANKS GLENN M	15324846	S SG	FOD
HARBISON ROBERT F JR	38176908	SGT	DNB
HUMPHREY BERNICE F	O-368485	1 LT	KIA
JOHNSON VERL B	38000542	S SG	KIA
KOBLER CORNELIUS M	O-666191	CAPT	M
MANCILLAS JOE L	38121800	PVT	DNB
MARQUEZ ELISEO	38011733	PVT	DNB
SAFFORD CHARLES V	O-890138	2 LT	KIA
THOMPSON MAX R	20759068	SGT	KIA

DNB = died non battle

25-86040

National Archives, WWII Army Casualties: New Mexico
https://www.archives.gov/research/military/ww2/army-casualties/new-mexico.html

The family was notified by mail of Clayton's death.

General Marshall

extends deep sympathy

in the loss of your son. He died

in the honorable service of his country

to preserve the freedom under which he lived

The Union County Leader newspaper published news of Clayton's death.

Mrs. Howell Receives News On Death Of Her Son

M|Sgt. Warren Whelchel of Amarillo, former member of the 200th div. A.A.C.A. and a prisoner of the Japanese, visited with Mrs. Mary Howell last week. He had worked with her son, Clayton F. Howell, on Bataan. He told Mrs. Howell that Clayton did not get a scratch during the war, went through the death march better than most as some of the men were sick or wounded and some died during and at the end of the march. He said that Clayton was one of the best liked of all the boys of the 200th and was a great encouragement to the others, and was certainly a good soldier. He said Clayton was in good health all the time he was in the Philippines, but took sick on the boat to Japan. He had several diseases but was recovering when he arrived in Japan, then took diphtheria and passed away. The boys were weakened on the trip as they only had crackers to eat the first 15 days, then the next 15 days they only had one meal a day.

Clayton Howell was 22 years of age, raised in Clayton and graduated from the Clayton high school with the class of 1937. He was well liked by all who knew him and a popular member among the younger class. His forefathers were true pioneers of New Mexico. Mrs. Howell is a first cousin to former Lt. Gov. Hiram Dow, and a granddaughter of Mrs. Susan Neatherlin, formerly Susan Roberts, sister to the late James E. Hinkle, former governor of New Mexico.

Mrs. Howell takes this means of expressing her thanks and appreciation to the many good friends for their wonderful encouragement and comfort during these years of suspense and grief.

They shall never be forgotten.

News of Clayton's death headlined the Union County Leader on February 10, 1943.

UNION COUNTY LEADER

CLAYTON	Largest Paid Up Circulation in the Trade Area	COUNTY SEAT
Population _____ 5,181		And Commercial Capital
Elevation _____ 00 feet	Center of Rich Livestock and Farming Country—Healthful Climate!	In Union County

VOLUME 15   CLAYTON, NEW MEXICO, THURSDAY.   February 10th   No. 50

Clayton Howell Dies While A Prisoner in Jap Concentration

Pfc. Clayton F. Howell, son of Mrs. Mary Howell of 417 Cherry Street, was reported last week by the War Department, to have died of disease in a Japanese prison camp. The War Department made public this announcement after receiving additional messages through the International Red Cross from Japan.

Clayton graduated from Clayton High School in 1937. On April 2, 1941 he enlisted in the army. In the fal lof 1941, he sailed for the Philippine Islands with the 200th Division, "New Mexico's Own" boys.

The last letter his parents received from him, on March 31, 1942, was sent out of Bataan by plane in February 1942. He said he was well and felt confident that the Japs would be whipped. The next word received was a telegram from the War Department ____ that Clayton was a prisoner of war of the Japanese.

Clayton was a great grandson of the late Susan Roberts Neatherlin, who was a pioneer of New Mexico and a sister of the former Governor Hinkle's wife. He is also a second cousin of Hiram Dow, former Lt. Governor of New Mexico.

The Leader joins the entire community in expressing sympathy to those who survive Clayton, who gave his life for a better world.

https://shorturl.at/5jzBG

Field Title	Value	Meaning
SERIAL NUMBER	38012614	38012614
NAME	HOWELL CLAYTON F	HOWELL CLAYTON F
GRADE, ALPHA	PFC	Private First Class
GRADE CODE	7	Second Lieutenant or Nurse or Dietitian or Physical therapy aide or Private First Class or Ensign or Second Class, Seaman
SERVICE CODE	1	ARMY
ARM OR SERVICE	CAC	Coast Artillery Corps
ARM OR SERVICE CODE	40	CAC: COAST ARTILLERY CORPS or AMP: ARMY MINE PLANTER SERVICE
DATE REPORT: DAY (DD)	07	07
DATE REPORT: MONTH (MM)	05	05
DATE REPORT: YEAR (Y)	2	1942
RACIAL GROUP CODE		
STATE OF RESIDENCE	83	New Mexico
TYPE OF ORGANIZATION		
PARENT UNIT NUMBER		
PARENT UNIT TYPE		
AREA	45	Southwest Pacific Theatre: Philippine Islands
LATEST REPORT DATE: DAY (DD)	17	17
LATEST REPORT DATE: MONTH (MM)	11	11
LATEST REPORT DATE: YEAR (Y)	2	1942
SOURCE OF REPORT	1	Individual has been reported through sources considered official.
STATUS	5	Died as Prisoner of War, Not Above Cases
DETAINING POWER	2	JAPAN
CAMP	709	Hoten POW Camp (Mukden) Manchuria 42-123
REP		
POW TRANSPORT SHIPS		

https://aad.archives.gov/aad/record-detail.jsp?dt=3159&mtch=1&cat=WR26&tf=F&sc=11675,11660,11679,11667,11669,11676,11672,11673&bc=,sl,fd&txt_11675=38012614&op_11675=0&nfo_11675=V,8,1900&rpp=10&pg=1&rid=121622

Few of Clayton's military files remain. According to the National Personnel Records Center (2025), his files were not located and likely destroyed in the fire of July 12, 1973.

Charmagne recalled a government official asking her mother if she wanted Clayton's remains returned home. Charmagne said her mother went to the train station to receive Clayton's remains, expecting a coffin but instead was handed a small box of his cremated remains. His remains are buried next to his parents in the Clayton Cemetery in Clayton, New Mexico.

Clayton's mother received a $10,000 military life insurance policy to help with burial expenses.

Veterans Administration
Insurance Form 1579

NOTICE OF SETTLEMENT

NATIONAL SERVICE LIFE INSURANCE

Date
July 7, 1944
NAME HOWELL,
Clayton F.
XC- 3,493,750

To Mrs. Mary Howell
417 Cherry Street
Clayton, New Mexico

You are hereby notified that, as a beneficiary of insurance

in the amount of $10,000.00   granted to   Clayton F. Howell

by the United States under the National Service Life Insurance Act of

October 8, 1940, as amended, you are entitled to monthly payments of $

$48.90   beginning November 17,   19 42 to continue for life.

The initial payment under this settlement will be dispatched to
you at the earliest possible date. If you should change your address,
the Accounting Division, Finance Service, Veterans Administration,
Washington 25, D. C., must be immediately notified.

All future communications with reference to this case must bear
the File Number XC-3,493,750.

Very truly yours,

H. L. McCOY,
Director of Insurance.

On May 23, 1946 Clayton's mother received a letter from one of Clayton's fellow soldiers. It read:

My Dear Mrs. Howell:

I received your letter as Mother forwarded it here where I am a patient. I am afraid I can't give you much information about Clayton. I knew him just a short while before he left the Philippines for Japan.

I am a doctor and I took care of him then, he had an attack of malaria, and as I recall he wasn't too sick then. I believe this illness was in Sept. of 1942. Soon after that he and his group sailed for Japan or Mukden, and I and my group were sent from Manila to Davao [City] Area. So much has happened since that I can't recall much of our acquaintanceship outside of saying that I knew him.

I knew the doctor in that camp tho, but I do not know his address, but his name is Capt. Elemer Shabart. He may be able to give you more information, but if the date is correct Clayton passed away soon after that camp was established.

I am very sorry that I am unable to give you any more information. But I wish you the best of luck and surely things will turn out smoothly for you.

Sincerely,
Calvin G. Jackson, Lt. Col. ME

May 23, '46

My dear Mrs. Howell:

I recieved your letter as
mother forwarded it to me here,
where I am a patient. I am
afraid I can't give you much
information about Clayton. I knew
him just a short while before
he left the Philippines for Japan.
I am a doctor and I took
care of him then, he had an
attack of malaria, and as I
recall he wasnt too sick then.
I believe this illness was in
Sept. of 1942. Soon after that he
and his group sailed for Japan
or Mukden, and I and my group
were sent from Manila to the
Ilanas area. So much has
happened since then that I cant
recall much of our acquaintanceship

outside of saying that I knew
him.

I knew the doctor in that
camp tho, but I do not know
his address, but his name is
Capt. Elmer Shabart (SHABART)
He may be able to give you
more information. But if the date
is correct Clayton passed away
soon after that camp was
established.

I am very sorry that I
am unable to give you any
more information. But I wish
you the best of luck and
surely things will turn out
smoothly for you.

Sincerely

Calvin G. Jackson
Lt. Col. M.C.

Lt. Col. Jackson wrote again on June 14, 1946 (Jackson, 1946).

My Dear Mrs. Clayton:

Thru my sister I got Capt. Shabart's address, and I'm hoping he can give you more information about Clayton as he was the doctor in that camp in Mukden.

Dr. E. J. Shabart
1427 Wicker Park Ave.
Chicago 22
Illinois

After leaving Bataan, Clayton was in O'Donnell Camp, until the early part of June '42. Then he was moved to Cabanatuan camp and was there until the last of October 42.

Sincerely,

Calvin G. Jackson

Thurs, 12, June

My Dear Mrs Clayton,

Thru my sister I got Capt Shabart's address, and I'm hoping he can give you more information about Clayton, as he was the doctor in that camp in Mukden.

Dr. E. J. Shabart,
1427 Wicker Park Ave,
Chicago 22,
Illinois

After leaving Bataan, Clayton was in O'Donnell Camp, until the early part of June '42. Then he was moved to Cabanatuan camp, & was there until the last of Oct. '42.

Sincerely
Calvin G. Jackson

Clayton's mother applied for a military headstone on September 26, 1949 and it was approved October 3, 1949.

On September 27, 1949, Clayton's mother received a request for reimbursement of interment or transportation expenses.

40904
GT-154-R

Payee

REQUEST FOR REIMBURSEMENT OF INTERMENT OR TRANSPORTATION EXPENSES W W I I
(Read Explanation on Reverse Side before completing form)

DATE: **27 Sept 1949**

NAME OF DECEDENT (Last, First, Middle Initial)
HOWELL, CLAYTON F.

BRANCH OF SERVICE
USAGF

TO BE FILLED IN BY CLAIMANT

A. ☒ INTERMENT EXPENSES
(Civilian or Private Cemetery)

RANK OR GRADE
1TC

SERIAL NO.
38012614

B. ☐ TRANSPORTATION EXPENSES
(National or Post Cemetery)

INSTRUCTIONS TO PERSONS SIGNING THIS FORM

1. This form is NOT to be signed by Funeral Director.
2. Fill in as required and sign four copies.
3. Check Box "A" or Box "B" above, not both.
4. Check Box "A" when interment is in a civilian or private cemetery.
5. Check Box "B" when remains are delivered to home or other place prior to burial in a national or post cemetery.

FILL IN THIS STATEMENT IF BOX "A" IS CHECKED	FILL IN THIS STATEMENT IF BOX "B" IS CHECKED
I certify that the sum of $ **75.00** was paid by me from personal funds in connection with the interment of the remains of the above-named decedent in the cemetery indicated below:	I certify that the sum of $ was paid by me from personal funds in connection with the transportation of the remains of the above-named decedent from: (City, town, or place from which remains were shipped)
NAME: Clayton Cemetery	
CITY OR COUNTY: Clayton	TO: (Name and Location of National or Post Cemetery)
STATE: New, Mexico	

RETURN FOUR COPIES TO
Commanding Officer
Distribution Center No. 13
Oakland Army Base
Oakland 14, California

Mary Howell

SIGNATURE OF CLAIMANT
Mary Howell
Clayton New Mex

ADDRESS (Street number or RFD, City and State)
620 Oak Street, Clayton, New Mexico

RELATIONSHIP TO DECEDENT
(Mother)

COPY

REMARKS

OCT 13 1949

QMC FORM 1236
REV 5 MAR 48

PREVIOUS EDITIONS OF THIS
FORM ARE OBSOLETE

16—54738-1

PART A

1. When the remains are delivered for interment in a civilian or private cemetery, you are responsible for paying all interment expenses. In this connection, you are entitled to the allowance mentioned in paragraph 2 below.

2. An amount not to exceed $75 is allowed by the Government toward actual interment expenses when final interment of the remains is in a private or civilian cemetery. No allowance is authorized toward interment expenses when interment is in a national or post cemetery.

3. The $75 maximum allowance by the Government toward interment expenses includes but is not limited to the payment of one or more of the following items: Hearse hire from the railroad station to your home, the funeral home, church, cemetery, or any other place designated by you; vault; church services; newspaper notices; transportation for friends and relatives to and from cemetery; and the services of a funeral director.

4. Reimbursement by the Government is made only to the person who paid from his personal funds the expenses of or incident to interment in a private or civilian cemetery. Receipted bills are not required to accompany this form. Any expenses over and above the $75 maximum must be borne by the person who incurred or paid the additional expenses.

PART B

1. When the remains are delivered to you at Government expense prior to burial in a national or post cemetery, you are responsible for all additional expenses necessary to deliver the remains from that point to the national or post cemetery grave site. However, you may be entitled to an allowance for the cost of transporting the remains from your home to the national or post cemetery grave site subject to the conditions outlined in paragraph 2 below.

2. Reimbursement of transportation expenses is allowed only when the cost to the Government to deliver the remains to you is LESS than what it would have cost the Government to deliver the remains direct to the national or post cemetery of final interment. However, the amount which you may be allowed (the difference between cost of delivery to you and cost of delivery by the Government direct to the national or post cemetery) may not exceed the amount actually expended by you to deliver the remains to the cemetery grave site. WHETHER OR NOT YOU WILL BE GRANTED AN ALLOWANCE IS DEPENDENT UPON AN AUDIT OF THIS REQUEST. IN ANY EVENT YOU WILL BE NOTIFIED OF ANY ALLOWANCE DUE YOU BY THE OFFICE TO WHICH THIS FORM IS SENT.

3. Reimbursement by the Government will be made only to the person who paid from his personal funds for transporting the remains to the national or post cemetery grave site.

4. No interment expense allowance is authorized since interment is made ultimately in a national or post cemetery.

U. S. GOVERNMENT PRINTING OFFICE   16—54738-1

Clayton's family received the Bataan Medal on his behalf. The inscriptions read: Bataan Medal. Awarded to the gallant New Mexico soldiers fighting for their country with the 200th Coast Artillery (AA) on December 7, 1941 by their state.

The hard-fought efforts of the brave soldiers will never be forgotten. "Of the 1800 men in the Regiment, less than 900 made it back home and, within one year, a third of them died from various complications. The 200th and 515th — The New Mexico Brigade — brought home four Presidential Unit Citations and the Philippine Presidential Citation. They earned their place in American History" (NM Bataan-Corregidor Memorial Foundation, n.d.b, para. 12 and 13).

Image: Wikipedia (n.d.b). Freely licensed media file repository.

NM Bataan and Corregidor Memorial Foundation unveiled a memorial in April, 2002 in Albuquerque, New Mexico to commemorate the New Mexico men who gave their lives in battles of Bataan and Corregidor and who were prisoners of war. Clayton's name is listed among his Union County comrades on slates of granite in the Bataan Memorial Park. Clayton's surviving sisters Bobbye, Charmagne, and Kay were invited to and attended the unveiling.

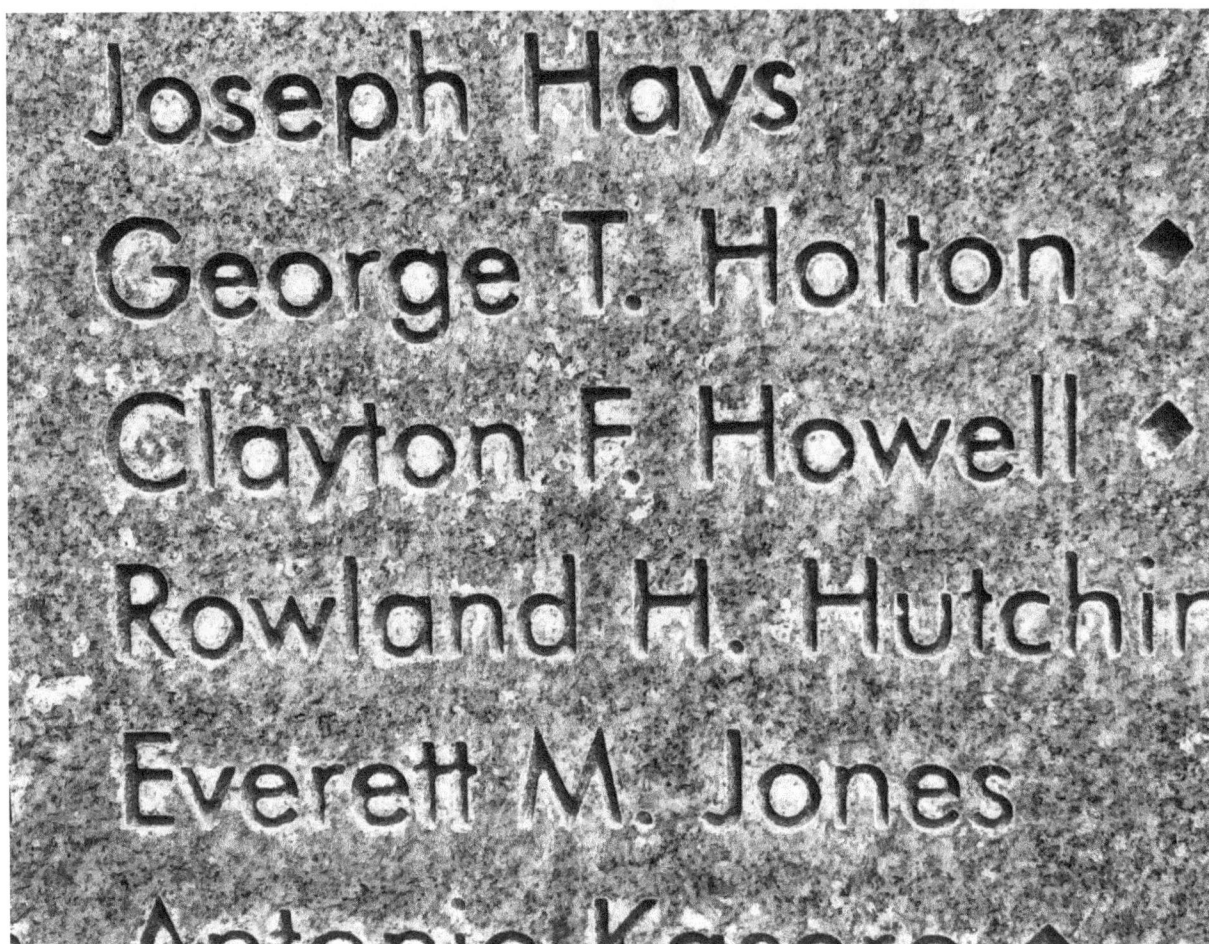

WWII Memorial

On April 29, 2004, the WWII Memorial was unveiled in Washington DC. The Memorial is managed by the National Parks Service.

Photo by Clayton's niece Rene Moore McCauley.

Clayton's sister Charmagne is pictured in the Bataan-Corregidor section of the Memorial. Seated in the scooter in the background is Charmagne's late husband, John Leroy Moore, a WWII veteran of the Navy. Photo by Clayton's niece Rene Moore McCauley.

REFERENCES

American Defenders of Bataan & Corregidor Memorial Society (n.d.).
https://www.adbcmemorialsociety.org/mukden-pows

Britannica (n.d.). Bataan Death March summary.
https://www.britannica.com/summary/Bataan-Death-March

FindAGrave (n.d.). Clayton Franklin Howell.
https://www.findagrave.com/memorial/6657266/clayton-franklin-howell

Griffin, M., ed. (n.d.). Photograph found with dead Japanese soldier. *Heroes of Bataan*. Marcus Griffin publisher.

Jackson, C. G. (1946, June 14). Personal communication.

Mansell, R. (2005). *Mukden (Hoten) timeline*. Citing Caples, C. B. (personal communication).
https://drive.google.com/drive/folders/1im2Nl1ssdbiW5eaG4mG4twmhxERGD6pY

McCauley, D. R. M. (n.d.). Photos of WWII Memorial.

Mobley, A. C. (1942). *To my soldier on Bataan*. Published on postcard.

National Archives (n.d.). WWII Army and Army Air Force casualties:
https://www.archives.gov/research/military/ww2/army-casualties

National Personnel Records Center (2025, March 5). Personal correspondence. National Archives.

NM Bataan-Corregidor Memorial Foundation (n.d.). *NM 200TH / 5l5TH: Honoring their story*.
https://www.bataanmemorialfoundationnm.org/

NM Bataan-Corregidor Memorial Foundation (n.d.b). *History: The battling bastards of Bataan*.
https://www.bataanmemorialfoundationnm.org/history

Union County Leader (1941-1942). Multiple newspaper articles. Terry Martin, publisher/editor.

Wikipedia (n.d.a). Camp O'Donnell. https://en.wikipedia.org/wiki/Camp_O%27Donnell

Wikipedia (n.d.b). *Presidential Unit Citation*. Freely licensed media file repository.
https://en.m.wikipedia.org/wiki/File:Presidential_Unit_Citation_(Philippines).svg

ABOUT THE AUTHOR

The author is Clayton's niece, Dana Kay Moore Gray, Ph.D. This book is made possible only because of the thoughtful care and preservation of artifacts and records by Clayton's family members including his mother and sister Margaret May "Bobbye" Howell Hoeme.

Dr. Gray is a semi-retired professor of education and marketing, a professor emeritus of Rogers State University, and a passionate genealogist. She is also a proud mom and Mimi. The eldest daughter of John Leroy Moore and Charmagne LaUna Howell, Dana grew up in the Kansas, Missouri, and Oklahoma region. A lifelong learner with a deep commitment to education, Dana earned her bachelor's degree from the University of Tulsa and completed her graduate studies at Oklahoma State University. She currently teaches doctoral students in education and MBA students. Before transitioning to academia, Dana spent more than 20 years in corporate marketing and public relations, during which she earned the Accredited Public Relations (APR) certification. Dana enjoys spending time with her family, traveling, and researching genealogy. She joins the millions of Americans who are forever indebted to the sacrifice and courage of our military veterans.

www.ingramcontent.com/pod-product-compliance
Lightning Source LLC
Chambersburg PA
CBHW080547090426
42734CB00016B/3226